The Wit and Wisdom of
Highland Cows

This is a STAR FIRE book

STAR FIRE BOOKS
Crabtree Hall, Crabtree Lane
Fulham, London SW6 6TY
United Kingdom

www.star-fire.co.uk

First published 2007

07 09 11 10 08

1 3 5 7 9 10 8 6 4 2

Star Fire is part of The Foundry Creative Media Company Limited



The CIP record for this book is available from the British Library.

ISBN: 978 1 84451 805 X

Printed in China

Thanks to: Cat Emslie, Andy Frostick, Sara Robson, Gemma Walters and Nick Wells

# *The Wit and Wisdom of* Highland Cows

Ulysses Brave

# *Foreword*

For years I studied Zen and the Art of
Animal Self-consciousness. Subsequently I
have written a large number of management,
self-help and philosophical texts over the
years, which have provided helpful advice
to those less fortunate than myself.
Here then, is my latest offering.

*Ulysses Brave*

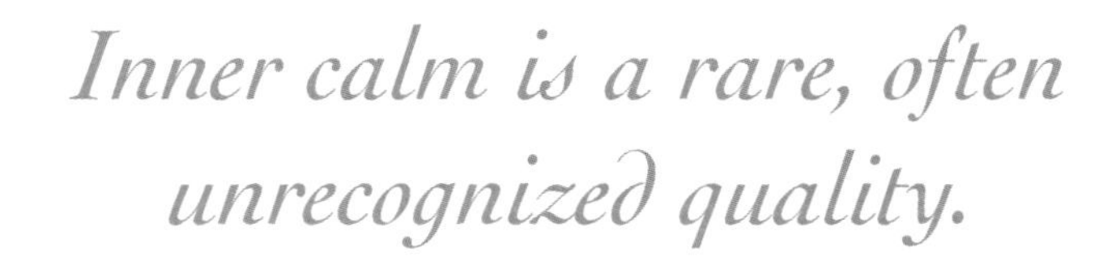
Inner calm is a rare, often
unrecognized quality.

*Celebrate your difference from the crowd, take pride in yourself.*

*Sometimes it seems as though everyone is against you. Try to find the reason before taking it personally.*

*Try to take the long view*
*of your life.*

*If you need to stand and fight, make sure you're ready to lose for your cause.*

*The invisible barriers which surround us all seem to appear when we least expect them. Don't give up! There may be other ways to your achieve your goals.*

*While the nurture versus nature debate still rages, remember to be kind and loving to your children and friends. Even if you have to whisper!*

*Don't forget to appreciate what you have.*

*Sometimes, the grass*
*really is greener on*
*the other side.*

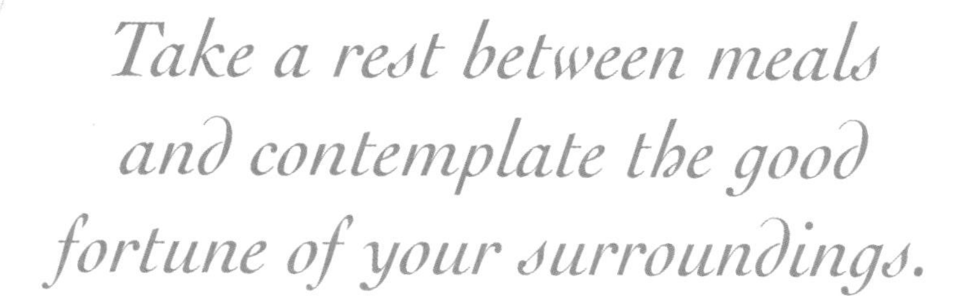

*Take a rest between meals and contemplate the good fortune of your surroundings.*

*If the tide goes out on your luck, make sure you're still standing in the same place, to benefit from its eventual return.*

*Zen-like one-eyed*
*looking can be useful*
*when trying to see beyond*
*the immediately obvious.*

*If fate seems bewildering, remember that your future is, at least in part, in your own hands.*

*Management manuals often refer to "decision drivers" and "drilling down". These can be useful tools for even the most basic tasks in life.*

*Romantic love can be hugely life-enhancing.*

*Try to perfect an ironic tilt of the head. Any angle between 3° and 5° will work well.*

*Caught between a rock and a hard place, always choose the rock. At least you will know what it looks like.*

*When dreaming it can be useful to observe your own multiple personalities of Self.*

*Try not to be*
*caught unawares.*

*When in repose, try to find ways of strengthening the muscles around your mouth. This will provide hours of entertainment in later life.*

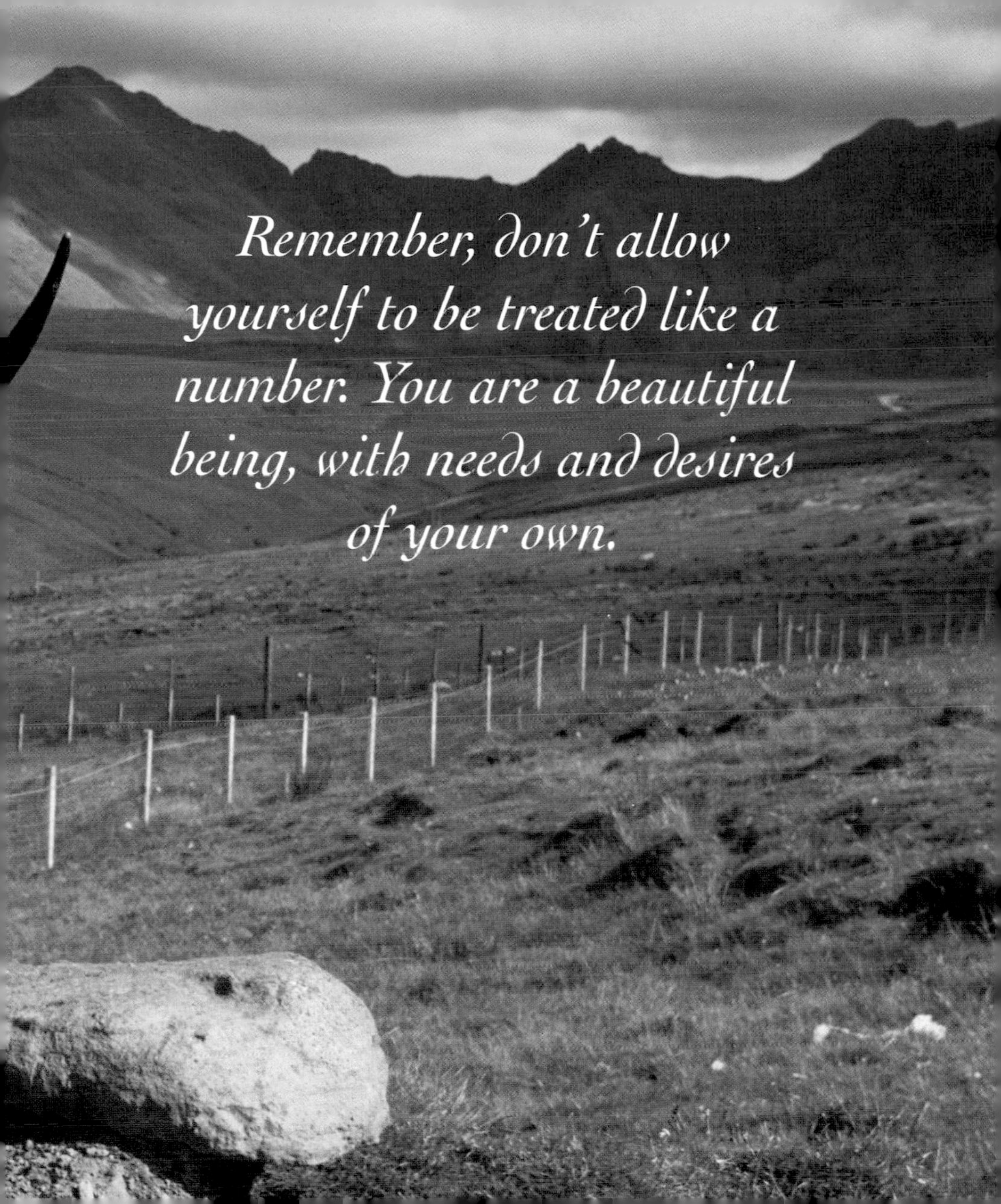

*Remember, don't allow yourself to be treated like a number. You are a beautiful being, with needs and desires of your own.*

*If you find yourself in a tight spot try reversing out slowly.*

*Try to break the vicious circle by stopping occasionally and rethinking your plan.*

*Sometimes our own shadow can come to life and offer support and guidance in difficult times.*

*If it all gets too much for you, try going to a quiet place and shouting as loud and long as you can.*

*Being leader of the pack isn't always about adulation and reward.*

Always lift your head high
and be proud of yourself.

*Try to look beyond your emotional barriers. Sometimes you have to leave behind friendships that were forged under different circumstances.*

*If you're stuck in a rut, try to lower your expectations to see life from a different perspective.*

*Sometimes burying your head to avoid a problem is the only option.*

*As we grow older, we behave in ways that were once regarded as unacceptable. Now we either don't know what we're doing, or we don't care!*

*Take care with exercise.*
*It can generate some*
*unpleasant symptoms.*

*Ah, just enjoy that early morning snack. You must fortify yourself for the busy day ahead.*

*See you soon ...*